# What Are Caves?

**by Mari C. Schuh**

**Consulting Editor: Gail Saunders-Smith, Ph.D.**

**Consultant: Sandra Mather, Ph.D., Professor Emerita, Department of Geology and Astronomy, West Chester University, West Chester, Pennsylvania**

Pebble Books

an imprint of Capstone Press
Mankato, Minnesota

Pebble Books are published by Capstone Press
1710 Roe Crest Drive, North Mankato, Minnesota 56003
www.capstonepub.com



*Library of Congress Cataloging-in-Publication Data*
Schuh, Mari C., 1975–
What are caves? / by Mari C. Schuh.
p. cm.—(Earth features)
Includes bibliographical references (p. 23) and index.
Summary: Simple text and photographs introduce caves and their features.
ISBN-13: 978-0-7368-1169-9 (hardcover)
ISBN-10: 0-7368-1169-9 (hardcover)
ISBN-13: 978-0-7368-4454-3 (softcover pbk.)
ISBN-10: 0-7368-4454-6 (softcover pbk.)
1. Caves—Juvenile literature. [1. Caves.] I. Title. II. Series.
GB601.2 .S37 2002
551.44′7—dc21 2001003764

# Note to Parents and Teachers

The Earth Features series supports national science standards for units on landforms of the earth. The series also supports geography standards for using maps and other geographic representations. This book describes and illustrates caves. The photographs support early readers in understanding the text. The repetition of words and phrases helps early readers learn new words. This book also introduces early readers to subject-specific vocabulary words, which are defined in the Words to Know section. Early readers may need assistance to read some words and to use the Table of Contents, Words to Know, Read More, Internet Sites, and Index/ Word List sections of the book.

Printed in China by Nordica.
0213/CA21300181
022013 007172R

# Table of Contents

A cave is a large, hollow space in the ground. Most caves are dark and damp.

Water creates most caves. Acid in water slowly wears away rock underground.

Some caves have
long, low tunnels.

Some caves have
tall, wide rooms.

entrance

twilight zone

dark zone

Caves have three zones.
They have an entrance,
a twilight zone,
and a dark zone.

Water in caves has calcium. Calcium builds up over time. It makes rock shapes called formations.

Stalactites are formations that hang from the top of a cave.

Stalagmites are formations that rise from the bottom of a cave.

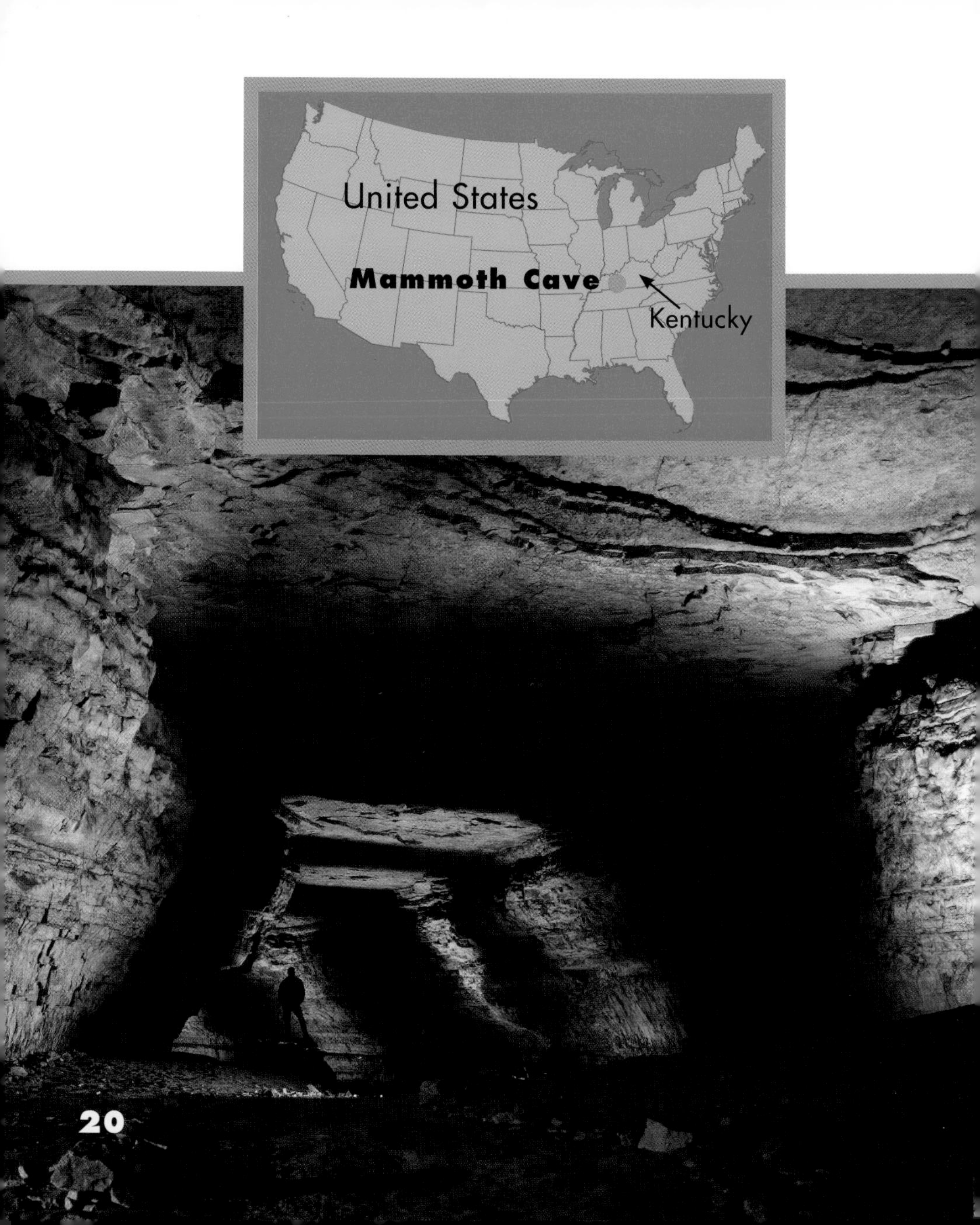
United States
Mammoth Cave
Kentucky

Mammoth Cave is
in Kentucky. It is the
longest group of caves
in the world.

# Words to Know

**acid**—a substance that sometimes is found in water; acid can wear away limestone in caves.

**calcium**—a mineral that can form stalactites, stalagmites, and other cave formations

**dark zone**—the part of a cave that is completely dark

**formation**—a rock in a cave that has an unusual shape

**Mammoth Cave**—a system of caves in western Kentucky

**stalactite**—a rock formation that hangs from the ceiling of a cave

**stalagmite**—a rock formation that rises from the floor of a cave

**twilight zone**—the part of a cave between the entrance and the dark zone

**zone**—an area that is separate from other areas

# Read More

**Brimner, Larry Dane.** *Caves.* A True Book. New York: Children's Press, 2000.

**Furgang, Kathy.** *Let's Take a Field Trip to a Cave.* Neighborhoods in Nature. New York: PowerKids Press, 2000.

**Llewellyn, Claire.** *Caves.* Geography Starts. Chigago: Heinemann Library, 2000.

# Internet Sites

FactHound offers a safe, fun way to find Internet sites related to this book. All of the sites on FactHound have been researched by our staff.

Here's all you do:

Visit *www.facthound.com*

**FactHound will fetch the best sites for you!**

# Index/Word List



**Word Count: 108**
**Early-Intervention Level: 16**

**Credits**
Kia Bielke, cover designer; Jennifer Schonborn, production designer and illustrator; Kimberly Danger, Mary Englar, and Jo Miller, photo researchers

Chip Clark, 10
Frederick D. Atwood, 16
Kevin Barry, 14
Laurence Parent Photography, 20
National Park Service, cover
Peter & Ann Bosted / TOM STACK & ASSOCIATES, 1
Robert & Linda Mitchell, 4, 18
Visuals Unlimited, 6; Albert J. Copley, 8